The Cell
A TINY CITY

Rebecca Woodbury, Ph.D., M.Ed.

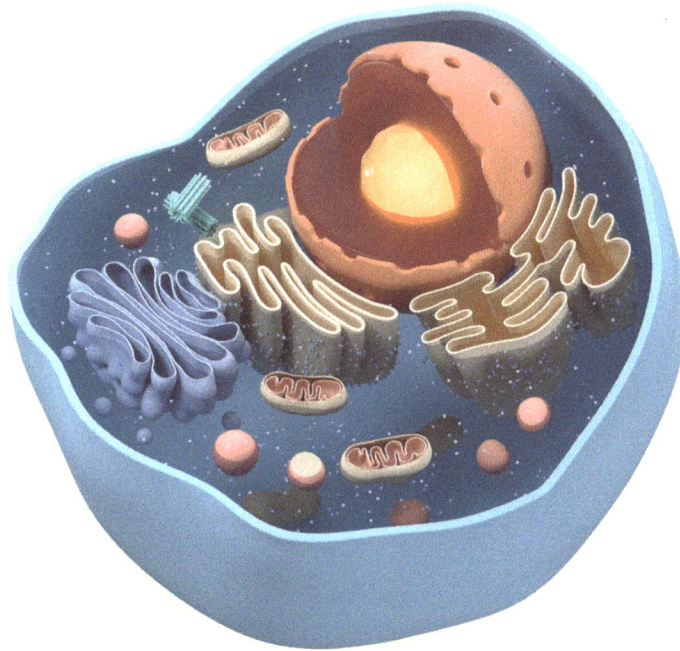

Gravitas Publications Inc.

The Cell
A TINY CITY

Illustrations: Janet Moneymaker

The Cell: A Tiny City
ISBN 978-1-950415-50-2

Published by Gravitas Publications Inc.
Imprint: Real Science-4-Kids
www.gravitaspublications.com
www.realscience4kids.com

RS4K

Photo credits: Cover and Title Page: By Vink Fan, AdobeStock; P. 4. By © 2017 Jee & Rani Nature Photography, License-CC BY-SA 4.0; P. 5. By Debby Hudson on Unsplash; P. 6. Frog, By Brian Gratwicke-CC BY SA 2.0; Ferns, By Marc Ryckaert, CC BY-SA 4-0; P. 7. Rocks, By Schorsch from Pixabay; Stars, By Baptiste Lheurette from Pixabay; P. 17. By Vink Fan, AdobeStock

Why is a frog a frog?

Why is a plant a plant?

Are you a plant?

Why are a frog and a plant...

...different from a rock and a star?

We know that
rocks, stars, frogs,
plants, and you...

...are all made of **atoms** and **molecules.**

Review: ATOMS

- **Atoms** are tiny building blocks that can link together.

- **Atoms** make everything we see, touch, taste, and smell.

Review: MOLECULES

Molecules are made

when **atoms link** together.

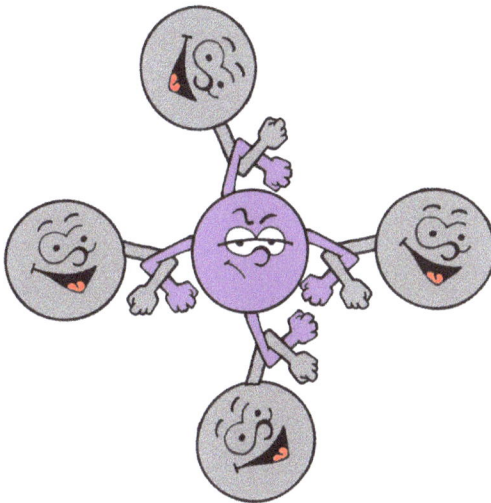

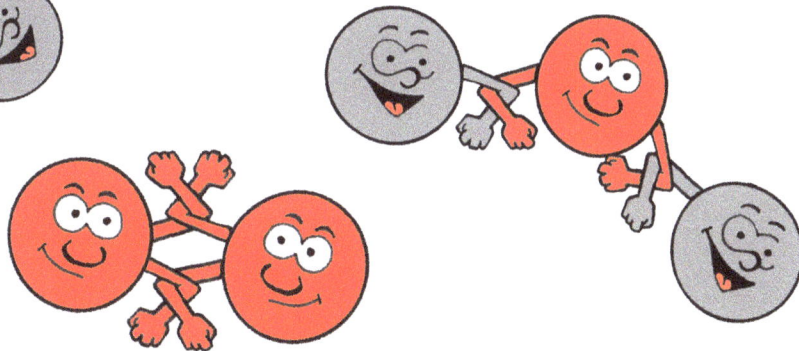

But frogs and plants are
different from rocks and stars.

WHY?

BECAUSE...

frogs and plants have **cells!**

Rocks and stars

do not have **cells.**

But wait!
What is a cell?

Cells are made of
atoms and molecules...

Atoms

Molecules

...**organized** into **proteins, DNA,**

and other special molecules.

A cell

Atoms and molecules are organized into DNA.

Atoms and molecules are organized into proteins.

Atoms and molecules are organized into a membrane that holds the cell.

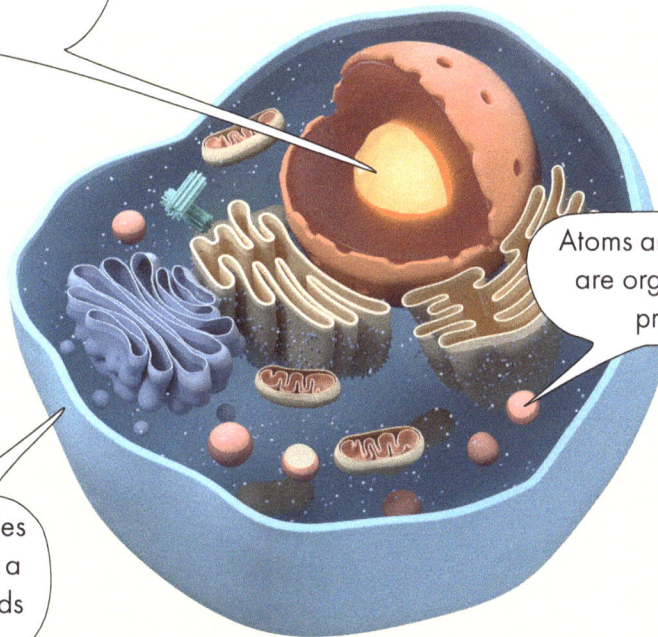

Review: PROTEINS

Proteins are molecules made of long strands of linked atoms.

Proteins fold into many different shapes and sizes.

Proteins do special jobs inside cells.

Review: DNA

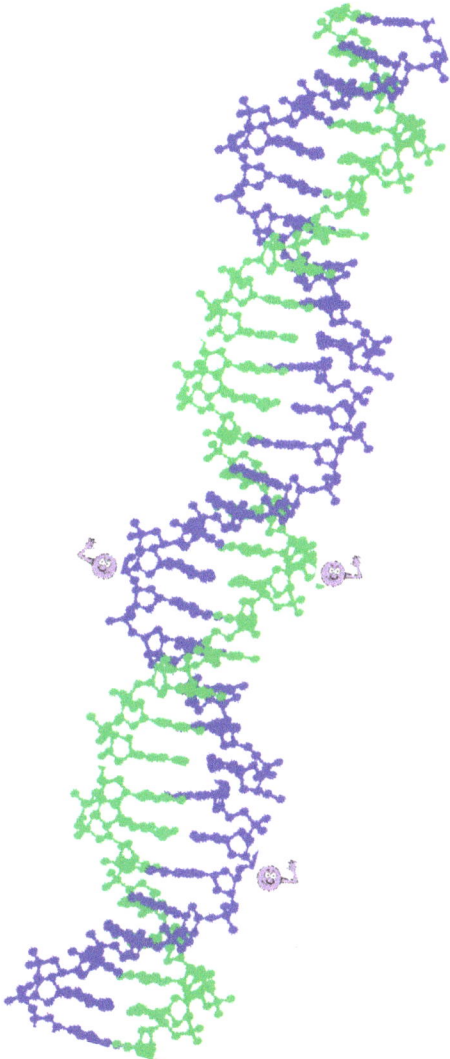

DNA stands for deoxyribonucleic acid.

DNA is a long strand of linked atoms.

The linked atoms make a special code.

This code tells cells what to do.

Proteins and DNA
work together like a
tiny cell city.

A Tiny Cell City

In a city people do different jobs.

Some people make food.

Some people move food.

Some people store food.

In a cell, similar things happen.

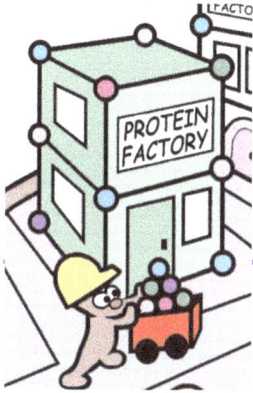

Proteins make food.

Proteins move food.

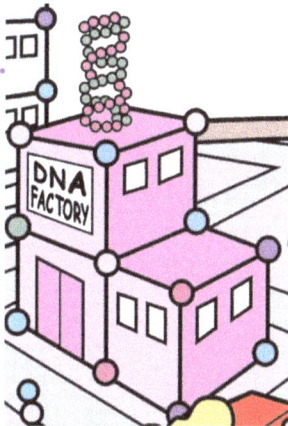

Proteins store food.

That is how cells make us alive!

How to say science words

atom   (AA-tum)

cell   (SEL)

deoxyribonucleic acid
(dee-AHK-see-riy-boh-new-klay-ik AA-suhd)

DNA   (DEE EN A)

living things   (LIH-ving THINGS)

membrane   (MEM-brayn)

molecule   (MAH-lih-kyool)

organize   (AWR-guh-niyz)

protein   (PROH-teen)

www.ingramcontent.com/pod-product-compliance
Lightning Source LLC
Chambersburg PA
CBHW040152200326
41520CB00028B/7582